SOUTH CAROLINA
TOTAL ECLIPSE GUIDE

Commemorative Official
Keepsake Guidebook

2017 Total Eclipse State Guide Series

Aaron Linsdau

SASTRUGI PRESS

JACKSON HOLE

Sastrugi Press / Published by arrangement with the author

South Carolina Total Eclipse Guide: Commemorative Official Keepsake Guidebook

Sastrugi Press
PO Box 1297, Jackson, WY 83001, United States
www.sastrugipress.com
Quantity sales: Special discounts are available on quantity purchases by corporations, associations, and others. For details, contact the publisher at the address above.

Library of Congress Catalog-in-Publication Data
Library of Congress Control Number: 2017905390
Linsdau, Aaron
South Carolina Total Eclipse Guide / Aaron Linsdau-1st United States edition
p. cm.
1. Nature 2. Astronomy 3. Travel 4. Photography
Summary: Learn everything you need to know about viewing, experiencing, and photographing the total eclipse in South Carolina on August 21, 2017.

ISBN-13: 978-1-944986-16-2
ISBN-10: 1-944986-16-2

508.4—dc23

Printed in the United States of America

All photography, maps and artwork by the author, except as noted.

10 9 8 7 6 5 4 3 2 1

Contents

Introduction

Thank you for purchasing this book. It has everything you need to know about the total eclipse in South Carolina on August 21, 2017.

A total eclipse passing across the United States is a rare event. The last US total eclipse was in 1979. It traveled over Washington, Oregon, Montana, and the corner of North Dakota.

The next total eclipse over the US will not be until April 8, 2024. It will pass over Texas, the Midwest, and on to Maine. After that, the next coast-to-coast total eclipse will be in 2045!

It's imperative to make travel plans today. You will be amazed at the number of people swarming to the total eclipse path. Some might say watching a partial versus a total eclipse is a similar experience. It's not.

This book is written for South Carolina visitors and anyone else viewing the eclipse. You will find general planning, viewing, and photography information inside. Should you travel to the eclipse path in South Carolina in mid-August, be prepared for an epic trip. The state estimates a half million visitors will converge on South Carolina.

Some hotels in the communities and cities along the path of totality in South Carolina have already sold out as of the writing of this guide. Finding lodging along the eclipse path will be a major challenge.

Resources will be stretched far beyond the normal limits. Think gas lines from the late 1970s. It may be likely that traffic along mountain highways will come to a complete standstill during this event. Be prepared with backup supplies.

Many smaller South Carolina towns are far from any major city. South Carolina country roads are slow. Please obey posted speed limits within all forest and park areas. Be cautious about believing a map application's estimate of travel time in South Carolina.

People in all communities along the path of the total eclipse plan to rent out their properties for this event. With a major celestial event in the summer of 2017, be assured that South Carolina "hasn't seen anything yet."

Is this to say to avoid South Carolina or other areas during the eclipse? Not at all! This guidebook provides ideas for interesting, alternative, and memorable locations to see the eclipse. It will be too late to rush to a better spot once the eclipse begins. Law enforcement will be out to help drivers reconsider speeding.

Please be patient and careful. There will be a large rush of people from all over the world converging on South Carolina to enjoy the total eclipse. Be mindful of other drivers on eclipse weekend, as they may not be familiar with South Carolina roads.

You should feel compelled to play hooky on August 21. Ask for the day off. Take your kids out of school. They'll likely be adults before the next chance to see a total eclipse. Create family memories that will last a lifetime. Sastrugi Press does not normally advocate skipping school or work. Make an exception because this is too big an event to miss.

Wherever you plan to be along the total eclipse path, leave early and remember your eclipse glasses. People from all around the planet will converge on South Carolina. Be good to your fellow humans and be safe. We all want to enjoy this spectacular show.

Visit www.sastrugipress.com/eclipse for the latest updates for this state eclipse book series.

Author Information

Polar explorer and motivational speaker Aaron Linsdau's first book, *Antarctic Tears*, is an emotional journey into the heart of Antarctica. He ate two sticks of butter every day to survive. Aaron coughed up blood early in the expedition and struggled through equipment failures. Despite the endless difficulties, he set a world record for surviving the longest solo expedition to the South Pole.

Aaron teaches audiences how the common person can achieve uncommon results. He shares his techniques for overcoming adrenaline burnout and constant overload. He inspires audiences to face their challenges with a new perspective. Aaron builds grit, teaches courage, and shows how to maintain a positive attitude in the face of adversity. He hopes that you will be inspired and have an enjoyable time watching the total eclipse in South Carolina.

Visit his websites: www.aaronlinsdau.com and www.ncexped.com

All About South Carolina

OVERVIEW OF SOUTH CAROLINA

As one of the original thirteen colonies, South Carolina has plenty history and culture to experience. The story of South Carolina is told partly through its national monuments and parks. The state has much to offer for those who come for a visit before or after the total solar eclipse during the summer of 2017.

Since South Carolina is in the southern region of the United States, the state can be warm year-round. During the summer, the temperature can average ninety degrees Fahrenheit. That coupled with the classic east coast humidity can make the state a warm environment. However, the lakes, rivers, and ocean coastline can help take the edge off the heat.

The state has been part of the United States' overall history since the beginning. The United States had to fight to build the country, and there have been countless battles fought on South Carolina's soil. As a consequence, many monuments, historic sites, and parks pay homage to those battles.

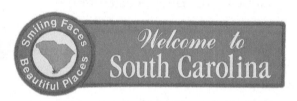

Both the national historic sites and national parks in South Carolina have significance to not only the state but also the country.

Fort Sumter was a Union military fort located on an island near beautiful Charleston, South Carolina. On April 12, 1861, the Civil War began with the Confederate army attacking the fort for thirty-four hours before the Union surrendered. Today Fort Sumter is maintained like it was back during the Civil War. People can come and visit the fort to learn more about its history, such as why the Confederates opened fire on the fort. This is the perfect site to begin an exploration of the South.

The town of Ninety-Six, South Carolina, is a site that features several battles from the American Revolution. The town got its name because that's how far it was from Charleston; Ninety-Six was a small, backcountry town. At this national historic site, people can experience life in the town, giving visitors a taste what life was like back in the 1700s. As a pivotal spot for the Revolutionary War, visitors can learn about the struggles the people of the town had to endure like being attacked by Cherokee Indians. Ninety-Six historic site is a good balance of experience and education, making it a great site to explore in South Carolina.

Kings Mountain Military Park is a national military park worth visiting. Located on the border of the Carolinas, the park is over four thousand acres, giving it a beautiful landscape for nature and animals. It is also the site of a battle that took place during the Revolutionary War. Thomas Jefferson stated that the battle "turned the tides" of the war, so the site holds its own historic importance. This park is perfect for learning about the battle and the Revolutionary War while spending time roaming and exploring.

Learning about the founding fathers is enjoyable and educational. With the musical sensation Hamilton playing in the country, people are starting to become interested in learning about more lesser-known founding fathers. Charles Pinckney would be one of those people. His house is located in Mount Pleasant just outside of Charleston, and it is now a national historic site. This site is dedicated to teaching about Pinckney and his importance in building the country, including his part in writing the Constitution and his political positions that he held. Much like Mount Vernon and Monticello, this site is perfect for further learning about an important figure in the country's history.

Congaree National Park is located in Hopkins southeast of Columbia. It is the largest hardwood forest in the southeast region of the country, giving it a large biodiversity of animals and plants. Along with the forest, the Congaree and Wateree Rivers run through the national park as well. The Congaree is a wonderful place to explore and experience a variety of wildlife.

The Overmountain Victory Trail is the trail that was used during the Revolutionary War for soldiers to reach the battle at Kings Mountain. This trail is over three hundred thirty miles, stretching over four states, one of them being South Carolina. If you enjoy a good challenge, it's possible to hike the entire trail in a few weeks. People can experience the hike that the men had to do before the battle. The Overmountain

Victory Trail is an excellent trail for a challenging and historical experience like no other.

South Carolina is home to countless historical and cultural sites to visit and learn from. Whether you visit these sites before or after the eclipse, there is no lack of locations to explore.

HOTELS AND MOTELS DURING THE ECLIPSE

Once word of the total eclipse over South Carolina spreads, rooms will become scarce. Many hotels in towns along the path of totality in western states have been sold out for a year or more. South Carolina is not alone in this challenge. Hotels in Idaho and Wyoming have been sold out for multiple years in anticipation of the total eclipse.

What does this mean for eclipse visitors? Lodging and room rentals in eclipse towns will be at a massive premium. Does that mean all hope is lost to find a place to stay? Not at all. But you will have to be creative. There will be few if any hotel rooms available in these eclipse cities by the time this book is printed. Accommodations in the cities and towns along the path of the eclipse have been sold out for months.

In spring 2017, the author searched on Hotels.com for rooms along the total eclipse path on the weekend of August 21 and found options still available. Once word of the eclipse spreads, room rates will increase and availability will drop.

Search for rooms farther away from the eclipse path. If you are willing to stay in cities outside the eclipse path, you will have better success at finding rooms. As the eclipse approaches, people will book rooms farther from the totality path. By midsummer, rooms in cities along the total eclipse path may be unavailable. The effect of this event will be felt across South Carolina and the rest of the United States.

Think regionally when looking for rooms. Be prepared to search far and wide during this major event. If a five-hour drive is manageable, your lodging options greatly expand, but it also increases your travel risk.

INTERNET RENTALS

To find rooms to stay in towns along the eclipse path, try a web service such as Airbnb.com. Note that many people rent out rooms or homes illegally, against zoning regulations. Many cities have already begun to feel the crunch of eclipse inquiries.

If South Carolina towns fully enforce zoning laws, authorities may prevent your weekend home rental. Online home rentals during the eclipse will be a target for rental scams. People from out of the area steal photos and descriptions, then post the home for rent. You send your check or wire money to a "rental agent" then show up to find you have been scammed. If the deal sounds strange or too good to be true, run away.

CAMPING

If you can book a campsite, do it now. Do not wait. All areas in the national forests are first-come, first-served. Forest roads will be packed. Expect all areas to be swarming with people. Show up early to stake out your spot. Consider staying farther away and driving early on August 21.

Please respect private land too. South Carolina folks don't take kindly to people overrunning their property without permission. In a big state with nearly five million residents, people are very protective, but they're friendly, too. You never know what you might be able to arrange with a smile and a bit of money.

This all said, there are plenty of camping opportunities throughout South Carolina. You don't have to sleep exactly on the eclipse path. If you're ready to rough it, there are national forest camping options.

Government agencies have been meeting since 2015 to talk about how to manage the influx of people. Every possible government agency will be working full time to enforce the various rules and regulations.

NATIONAL PARKS AND MONUMENTS

Chances are finding a camping site at any state park, national park, or national monument in South Carolina will be challenging. To watch the eclipse from any location, you do not have to sleep in it. You just need to drive there in the morning.

Law enforcement will be present on the eclipse weekend. Hundreds of thousands of people are expected in the region. Parking will overflow. It will make parking lots and lines on Black Friday at the mall look uncrowded. For an event of this magnitude, find parking early.

The first sentence of the national parks mission statement is:

"The National Park Service preserves unimpaired the natural and cultural resources and values of the national park system for the enjoyment, education, and inspiration of this and future generations."

Roadside camping (sleeping in your car) is not allowed in national monuments or parks. Park facilities are only designed to handle so many people per day. Water, trash collection, and toilets can only withstand so much. If you notice trash on the ground, take a moment to throw it away. Protect your national park and help out. Rangers are diligent and hardworking but they can only do so much to manage the expected crowds.

NATIONAL FORESTS AND WILDERNESS

There are national forest options in South Carolina. They all have camping opportunities. The forest service manages undeveloped and primitive campsites. Be sure to check for any fire restrictions. Check with individual agencies for last-minute information and regulations. The forest service requires proper food storage. Plan to purchase food and water before choosing your campsite. Below is a partial list of national forests and wilderness areas to visit in the state:

Francis Marion and Sumter National Forests:
   www.fs.usda.gov/scnfs
Ellicott Rock Wilderness:
   www.fs.usda.gov/recarea/scnfs/recarea/?recid=47089
Hell Hole Swamp

Backcountry service roads abound in South Carolina. Maps for forests are available at local visitor centers and bookstores. This book's website has digital copies of some forest maps.

Printed national forest maps are large and detailed. They have il-lustrated road paths, connections, and other vital travel information not available on digital device maps. Viewing digital maps on your smartphone or iPad is difficult. If you plan to camp in the forest, a real paper map is a wise investment.

Camping in federal wilderness areas is also allowed. Those areas afford the ultimate backcountry experience. However, be aware that no vehicle travel is allowed in the specially designated areas. This ban includes: vehicles, bikes, hang gliders, and drones. You can travel only on foot or with pack animals.

SLEEP IN YOUR CAR

Countless RVs, campers, trucks, cars, and motorcycles will flood South Carolina. Sleeping in your car with friends is tolerable. Doing so with unadventurous spouses or children is another matter.

Do not be caught along the path of the total eclipse without some sort of plan, especially in the bigger cities of South Carolina. The whole path of totality will fill with people on August 21.

USEFUL LOCAL WEBCAMS

Local webcams are handy to make last-minute travel decisions. The webcams are sensitive enough to show headlights at night. Use them to determine if there are issues before traveling out. Eclipse traffic will add to the morning commuter traffic.

The smartphone application Wunderground is useful to check on webcams in one place. All the webcams are listed in the app.

Weather

It's all about the weather during the eclipse. Nothing else will matter if the sky is cloudy. You can be nearly anywhere in South Carolina and catch a view of the sky when traffic comes to a standstill. But if there's a cloud cover forecast, seriously reconsider your viewing location.

Travel early wherever you plan to go. Attempting to change locations an hour before the eclipse due to weather will likely cause you to miss the event. South Carolina country roads can be narrow and slow. The number of vehicles will cause unexpected backups.

MODERN FORECASTS

Use a smartphone application to check the up-to-date weather. Wunderground is a good application and has relatively reliable forecasts for the region. The hourly forecast for the same day has been rather accurate for the last two years. The below discussion refers to features found in the Wunderground app. However, any application with detailed weather views will improve your eclipse forecasting skills.

CLOUD COVER FORECAST

The most useful forecast view is the visible and infrared cloud-coverage map. Avoid downloading this app the night before and trying to

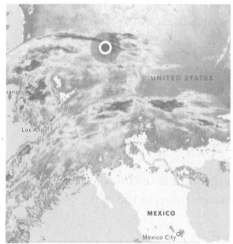

Infrared cloud map showing the worst case eclipse cloud cover.
Courtesy of National Weather Service.

learn how to read it. Practice reading them at home. It's imperative to understand how to interpret the maps early.

All cloud cover, night or day, will appear on an infrared map. Warm, low-altitude clouds are shown in white and gray. High-altitude cold clouds are displayed in shades of green, yellow, red, and purple. Anything other than a clear map spells eclipse-viewing problems.

To improve your weather guess, use the animated viewer of the cloud cover. It will give you a sense of cloud motion. You can discern whether clouds or rain are moving toward, away from, or circulating around your location.

NORMAL SOUTH CAROLINA WEATHER PATTERN

Due to the direction of the jet stream, most weather travels across the Pacific Ocean, through the western states, over Tennessee and North Carolina, and into South Carolina. On occasion, weather can approach from any direction. Due to the nature of the tropical storms from the Atlantic, weather in South Carolina can be unpredictable.

The common weather pattern in August is hot and humid. Passing cold fronts in summer can bring unexpected cloud cover.

Cities in South Carolina tend to have variable clouds during August. Prepare to make adjustments. If anything other than clear skies are predicted, plan to drive to other parts of Tennessee, South Carolina, North Carolina, or Georgia.

Be aware of tornadoes in South Carolina. Although the peak tornado season is June, there have been many recorded tornadoes in August. Pay attention to the weather forecast. If dangerous weather is predicted, your main concern should be safety rather than chasing an eclipse.

Consider that slow-moving clouds can obscure the sun for far longer than the two-minute duration of the totality. The time of totality is so short that you do not want to risk it. Missing it due to a single cloud will be a major disappointment.

LOCAL ECLIPSE WEATHER FORECASTS

Local town and city newspapers, radio, and television stations around South Carolina will have a weekend edition with articles discussing the eclipse weather. However, conditions change unpredictably in South Carolina. A three-day forecast may be incorrect.

FOREST FIRES

For the past several years, forest fires have been common in the United States. The summer of 2017 is likely to be no different. There were fires in the forests South Carolina during spring 2017. Chances are there will be fires again in the region in the summer of 2017. For fire updates check:

inciweb.nwcg.gov

For the best eclipse viewing experience, you need to have as clear a sky as possible. Fog, clouds, or smoke will

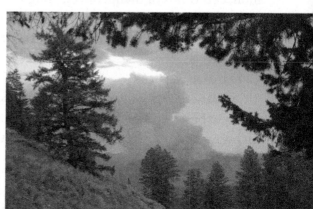

obscure the subtleties of the sun's corona. If you think the view of the sky is going to be blocked, don't wait until the last minute to move to a clearer location. If you wait too long to decide to move to a better viewing area, it may be impossible due to traffic.

ROAD CLOSURES DUE TO FIRES

Highways connecting various South Carolina towns can be closed during a major fire. The Pinnacle Mountain Fire scorched nearly ten-thousand acres and blackened South Carolina skies. There were other fires in the region at the same time, too.

With unpredictable weather in the last few years, it's a guess what will happen in August 2017. If forest areas continue to remain dry, the whole region may have many fires.

Although fire is an important part of forest ecology, it worries eclipse chasers. Other than clouds, smoke from fires will block the view of the sun and moon on the morning of the eclipse. Should there be fires where you are or may be headed, reconsider your location as early as possible. The most accurate website for fires is:

inciweb.nwcg.gov

Check the South Carolina road report for updated information:

www.511sc.org

It's imperative to plan for fires and their effects. Watch the weather reports. If strong winds and lightning storms are forecast, prepare to change your viewing location. If conditions are poor, you and thousands of other vehicles will be trapped in slow-moving traffic.

If you believe it's necessary to leave a town to watch the eclipse, do so the night before or extremely early in the morning. RVs are common, and trains of them crawl through popular areas.

South Carolina Information

CELLULAR PHONES

Cellular "cell" phone service in remote South Carolina locations is spotty at best. Most of the time there is good coverage along the main highways and interstates. However, even along major thoroughfares, there can be little or no coverage.

It's possible to find zones where text messages will send when phone calls are impossible. If you cannot make a phone call, the chance of having data coverage for web surfing or e-mail is low.

Please look up any information or communicate what you need before departing from the main roads around South Carolina. Bureau of Land Management (BLM) areas sometimes have coverage. Planned to be self-contained. Treat like your cell phone like it won't connect.

You may find yourself out of cell service. With a large number of cell users in a concentrated area, coverage and data speed may collapse as well. Search on the phrase "cell phone coverage breathing".

Wilderness and Forest Safety

All South Carolina forests are full of wild animals. Although beautiful, wild animals can be dangerous. They can easily injure or kill people, as they are far more powerful than humans. Do not try to feed any wild animals, including squirrels, foxes, and chipmunks, as they can carry diseases. These suggestions apply to all public lands.

BEARS

The forests of South Carolina are home to black bears. Safety is imperative around these powerful animals. Although they often appear docile, they can become aggressive if threatened. In the unlikely event of an attack, fight back against the bear. Use whatever you have at your disposal to defend yourself. Report all negative or aggressive bears to the local authorities.

If a bear hears you, it will usually vacate the area. Bear charges are often caused by unexpected and surprise encounters. Noise is the best defense to avoid surprising bears. Regularly clap, make noise, and talk loudly. The South Carolina Department of Natural Resources has specific information on safety and food management in bear country at www.dnr.sc.gov/wildlife/publications/nuisance/blackbears.pdf.

It is recommended to stay one hundred yards (300 feet) away from all bears. They are exciting to see but need their space. Refer to current forest or park regulations for more safety information.

RATTLESNAKES

There are several primary species of rattlesnakes in South Carolina: Copperhead, Coral, Cottonmouth, Pygmy Rattlesnake, Eastern Diamondback, and Timber Rattlesnakes. Although these reptiles are not generally aggressive, they can strike when provoked or threatened. Of the approximately 8,000 people annually bitten by venomous snakes in the United States, ten to fifteen people die according to the U.S. Food and Drug Administration.

The best way to avoid rattlesnake encounters is to be mindful of your environment. Do not place your hands or feet in locations where you cannot clearly see the surroundings. Avoid heavy brush or tall weeds where snakes hide during the day. Step on a log or rock rather than over it, as a hidden snake might be on the other side. Rattlesnakes may not make any noise before striking.

Avoid handling all rattlesnakes. Should you be bitten, stay calm and call 911 or emergency dispatch as soon as possible. Transport the victim to the nearest medical facility immediately. Rapid professional treatment is the best way to manage rattlesnake bites. Refer to US Forest Service and professional medical texts for more information on managing rattlesnakes injuries.

TICKS

Ticks exist all across the United States but not all species transmit disease. Ticks cannot fly or jump, but they climb grasses in shrubs in order to attach to people or animals that pass by. Ticks feed on the blood of their host. In doing so, they can transmit potentially life-threatening diseases such as Lyme disease.

According to the National Pesticide Information Center, ticks must be attached and feed for several hours before an infection can be passed. Do not wait that long to manage tick exposure. As soon as you travel through outdoor locations, especially in grasses or shrubs, check yourself for ticks. They can be incredibly small and difficult to detect. Some are smaller than the head of a pin.

You may wish to consider wearing tick-specific insect repellent. Check with your doctor or medical professional about any potential adverse side effects of chemical repellents.

Should you discover an attached tick, follow modern tick removal methods. There are many incorrect tick removal methods found online. Refer to your medical professional for proper removal methods. Improper removal increases the likelihood of an infection.

MOUNTAIN LIONS

Though listed as extinct, there have been recent reports of mountain lions in South Carolina. If you encounter a mountain lion, do not run. Keep calm, back away slowly, and maintain eye contact. Do all you can to appear larger. Stand upright, raise your arms, or hoist your jacket. Never bend over or crouch down. If attacked, fight back.

Eclipse Day Safety

1. Hydrate
Summers can be extremely warm. The excitement of the event can distract you from managing hydration. Drink plenty of water. Consume more than you would at home.

2. Eye Safety time
Use certified eclipse safety glasses at all times when viewing the partial eclipse. Only remove the glasses when the totality happens. Give your eyes time to rest. Even though the glasses are safe, your eyes can dry out and become irritated. Bring FDA approved eye drops to keep your eyes moist.

3. Sun exposure
The sun is much more intense in August. Wear sunglasses and liberally apply sunscreen to avoid sunburns.

4. Eat well
Keep your energy up. Appetite loss is common when traveling. Maintain your normal eating schedule.

5. Prepare for temperature changes
Temperatures will drop rapidly once the sun sets, especially in the mountains or forests. Bring appropriate clothing.

6. Talk with your doctor
If the humidity or heat bothers you, talk with your doctor before traveling. Seek professional medical attention for serious symptoms.

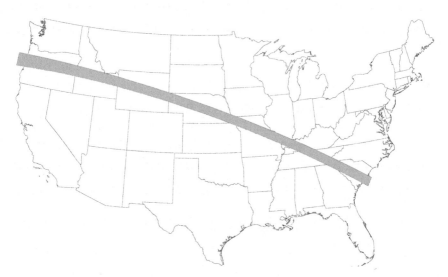

Total eclipse path across the United States (approximate).

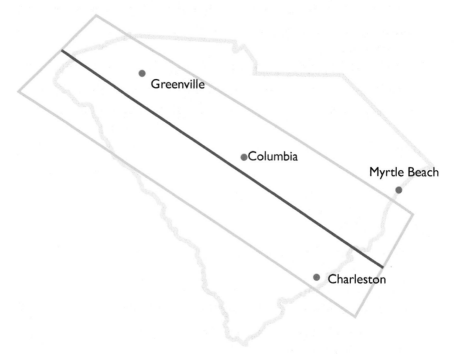

Total eclipse path across South Carolina (approximate).

All About Eclipses

HOW AN ECLIPSE HAPPENS

An eclipse occurs when one celestial body falls in line with another, thus obscuring the sun from view. This occurs much more often than you'd think, considering how many bodies there are in the solar system. For instance, there are over 150 moons in the solar system. On Earth, we have two primary celestial bodies: the sun and the moon. The entire solar system is constantly in motion, with planets orbiting the sun and moons orbiting the planets. These celestial bodies often come into alignment. When these alignments cause the sun to be blocked, it is called an eclipse.

For an eclipse to occur, the sun, Earth, and moon must be in alignment. There are two types of eclipses: solar and lunar. A solar eclipse occurs when the moon obscures the sun. A lunar eclipse occurs when the moon passes through Earth's shadow. Solar eclipses are much more common, as we experience an average of 240 solar eclipses a century compared to an average of 150 lunar eclipses. Despite this, we are more likely to see a lunar eclipse than a solar eclipse. This is due to the visibility of each.

For a solar eclipse to be visible, you have to be in the moon's shadow. The problem with viewing a total eclipse is that the moon casts a small shadow over the world at any given time. You have to be in

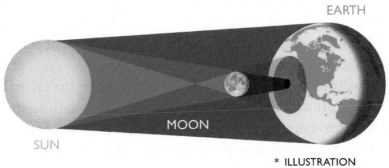

EARTH

MOON

SUN

* ILLUSTRATION NOT TO SCALE

a precise location to view a total eclipse. The issue that arises is that most of these locations are inaccessible to most people. Though many would like to see a total solar eclipse, most aren't about to set sail for the middle of the Pacific Ocean. In fact, a solar eclipse is visible in the same place on the world on average every 375 years. This means that if you miss a solar eclipse above your hometown, you're not going to see another one unless you travel or move.

It's much easier to catch a glimpse of a lunar eclipse, even though they occur at a much lower frequency than their solar counterparts. A lunar eclipse darkens the moon for a few hours. This is different than a new moon when it faces away from the sun. During these eclipses, the moon fades and becomes nearly invisible.

Another result of a lunar eclipse is a blood moon. Earth's atmosphere bends a small amount of sunlight onto the moon turning it orange-red. The blood moon is caused by the dawn or dusk light being refracted onto the moon during an eclipse.

Lunar eclipses are much easier to see. Even when the moon is in the shadow of Earth, it's still visible throughout the world because of how much smaller it is than Earth.

TOTAL VS. PARTIAL ECLIPSE

What is the difference between a partial and total eclipse? A total eclipse of either the sun or the moon will occur only when the sun, Earth, and the moon are aligned in a perfectly straight line. This ensures that either the sun or the moon is partially or completely obscured.

In contrast, a partial eclipse occurs when the alignment of the three celestial bodies is not in a perfectly straight line. These types of eclipses usually result in only a part of either the sun or the moon being obscured. This is often what led to ancient civilizations believing that some form of magical beast or deity was eating the sun or the moon. It appears as though something has taken a bite out of either the sun or the moon during a partial eclipse.

Total eclipses, rarer than partial eclipses, still occur quite often. It's more difficult for people to be in a position to experience such an event firsthand. Total solar eclipses can only be viewed from a small portion of the world that falls into the darkest part of the moon's shadow. Often this happens in the middle of the ocean.

The Moon's Shadow

The moon's shadow is divided into two parts: the umbra and the penumbra. The former is much smaller than the latter, as the umbra is the innermost and darkest part of the shadow. The umbra is thus the central point of the moon's shadow, meaning that it is extremely small in comparison to the entire shadow. For a total solar eclipse to be visible, you need to be directly beneath the umbra of the moon's shadow. This is because that is the only point at which the moon completely blocks the view of the sun.

In contrast, the penumbra is the region of the moon's shadow in which only a portion of the light cast by the sun is obscured. When

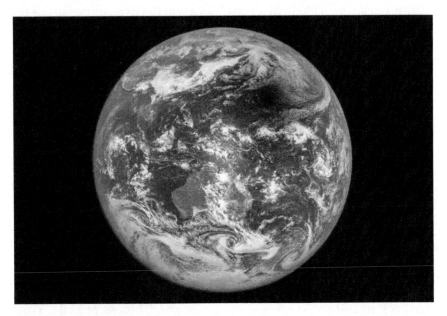

Total eclipse shadow 2016 as seen from 1 million miles on the Deep Space Climate Observatory satellite. Courtesy of NASA.

standing in the penumbra, you are viewing the eclipse at an angle. In the penumbra, the moon does not completely block the sun from view. This means that while the event is a total solar eclipse, you'll only see a partial eclipse. The umbra for the August 21st eclipse is approximately sixty miles wide. The penumbra will cover much of the United States.

To provide some context, the last total solar eclipse we experienced occurred on March 9, 2016, and was visible as a partial eclipse across most of the Pacific Ocean, parts of Asia, and Australia. However, the only place in the world to view this total solar eclipse was in a few parts of Indonesia.

Due to the varied locations and the brief periods for which they're visible, it's difficult to see each and every eclipse that occurs. Many people don't even realize that they have occurred. Consider that the umbra of the moon represents such a small fraction of the entire shadow and the majority of our planet is comprised of water. Thus, the rarity of being able to view a total solar eclipse increases significantly because it's likely that the umbra will fall over some part of the ocean rather than a populated landmass.

ECLIPSES THROUGHOUT HISTORY

Ancient peoples believed eclipses were from the wrath of angry gods, portents of doom and misfortune, or wars between celestial beings. Eclipses have played many roles in cultures, creating myths since the dawn of time. Both solar and lunar eclipses affected societies worldwide. Inspiring fear, curiosity, and the creation of legends, eclipses have cast a long shadow in the collective unconscious of humanity throughout history.

EARLY MYTH & ASTRONOMY

Documented observations of solar eclipses have been found as far back in history as ancient Egyptian and Chinese records. Time-keeping was important to ancient Chinese cultures. Astronomical observations were an integral factor in the Chinese calendar. The first observation of a solar eclipse is found in Chinese records from over 4,000 years ago. Evidence suggests that ancient Egyptian observations may predate those archaic writings.

Many ancient societies, including Roman, Greek and Chinese civilizations, were able to infer and foresee solar eclipses from astronomical data. The sudden and unpredictable nature of solar eclipses had a stressful and intimidating effect on many societies that lacked the scientific insight to accurately predict astronomical events. Relying on the sun for their agricultural livelihood, those societies interpreted solar eclipses as world-threatening disasters.

In ancient Vietnam, solar eclipses were explained as a giant frog eating the sun. The peasantry of ancient Greece believed that an eclipse was the sign of a furious godhead, presenting an omen of wrathful retribution in the form of natural disasters. Other cultures were less speculative in their investigations. The Chinese Song Dynasty scientist Shen Kuo proved the spherical nature of the Earth and heavenly bodies through scientific insight gained by the study of eclipses.

THE ECLIPSE IN NATIVE AMERICAN MYTHOLOGY

Eclipses have played a significant role in the history of the United States. Before Europeans settled in the Americas, solar eclipses were important astronomical events to Native American cultures. In most native cultures, an eclipse was a particularly bad omen. Both the sun and the moon were regarded as sacred. Viewing an eclipse, or even being outside for the duration of the event, was considered highly taboo by the Navajo culture. During an eclipse, men and women would simply avert their eyes from the sky, acting as though it was not happening.

The Choctaw people had a unique story to explain solar eclipses. Considering the event as the mischievous actions of a black squirrel and its attempt to eat the sun, the Choctaw people would do their best to scare away the cosmic squirrel by making as much noise as possible until the end of the event, at which point cognitive bias would cause them to believe they'd once again averted disaster on an interplanetary scale.

CONTEMPORARY AMERICAN SOLAR PHENOMENA

The investigation of solar phenomena in twentieth-century American history had a similarly profound effect on the people of the United States. A total solar eclipse occurring on the sixteenth of June, 1806, engulfed the entire country. It started near modern-day Arizona. It passed across the Midwest, over Ohio, Pennsylvania, New York, Massachusetts, and Connecticut. The 1806 total eclipse was notable for being one of the first publicly advertised solar events. The public was informed beforehand of the astronomical curiosity through a pamphlet written by Andrew Newell entitled *Darkness at Noon, or the Great Solar Eclipse.*

This pamphlet described local circumstances and went into great detail explaining the true nature of the phenomenon, dispelling myth and superstition, and even giving questionable advice on the best methods of viewing the sun during the event. Replete with a short historical record of eclipses through the ages, the *Darkness at Noon* pamphlet is one of the first examples of an attempt to capitalize on the mysterious nature of solar eclipses.

Another notable American solar eclipse occurred on June 8, 1918. Passing over the United States from Washington to Florida, the eclipse was accurately predicted by the U.S. Naval Observatory and heavily documented in the newspapers of the day. Howard Russell Butler, painter and founder of the American Fine Arts Society, painted the eclipse from the U.S. Naval Observatory, immortalizing the event in *The Oregon Eclipse.*

Four more total solar eclipses occurred over the United States in the years 1923, 1925, 1932, and 1954, with another occurring in 1959. The October 2, 1959, solar eclipse began over Boston, Massachusetts. It was a sunrise event that was unviewable from the ground level. Eminent astronomer Jay Pasachoff attributed this event to sparking his interest in the study of astronomy. Studying under Professor Donald Menzel of Williams College, Pasachoff was able to view the event from an airline hired by his professor.

To this day, many myths surround the eclipse. In India, some local customs require fasting. In eastern Africa, eclipses are seen as a danger to pregnant women and young children. Despite the mystery and legend associated with unique and rare astronomical events, eclipses continue to be awe-inspiring. Even in the modern day, eclipses draw out reverential respect for the inexorable passing of celestial bodies. They are a reminder of the intimate relationship between the denizens of Earth and the universe at large.

PRESENT DAY ECLIPSES

The year 2016 brought the world just two solar eclipses. A total solar eclipse occurred on the 9th of March. An annular solar eclipse, in which the sun appears as a "ring of fire" occurred on the 23rd of March. If you're interested in seeing this rare and exciting solar

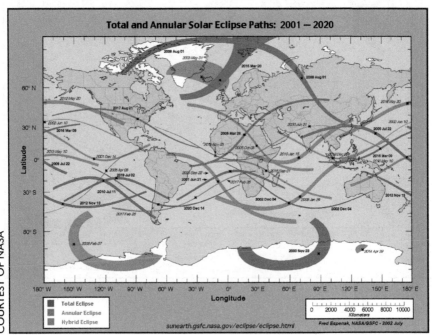

phenomenon yourself, you must to travel to either South America or Western Africa on the 26th of February, 2017.

The next total solar eclipse viewable from the United States, or anywhere else in the world, will occur on the 21st of August, 2017. It will be visible in Oregon, Idaho, Wyoming, Nebraska, Kansas, Missouri, Illinois, Kentucky, Tennessee, Georgia, North Carolina, and South Carolina. The event will be the only total solar eclipse for Americans this decade.

FUTURE AMERICAN ECLIPSES

The next total eclipse to cross the continental United States is on April 8, 2024. It will travel from Texas to Maine. After that, the next American total eclipses will be in 2044 and 2045.

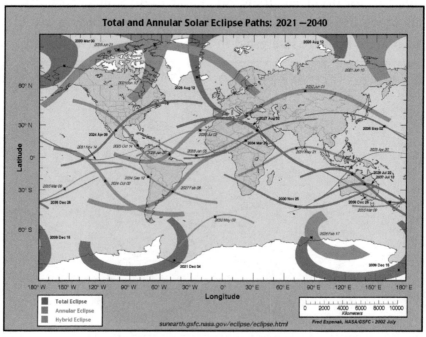

COURTESY OF NASA

Viewing and Photographing the Eclipse

At-home Pinhole Method

Use the pinhole method to view the eclipse safely. It costs little but is the safest technique there is. Take a stiff piece of single-layer cardboard and punch a clean pinhole. Let the sun shine through the pinhole onto another piece of cardboard. That's it!

Never look at the sun through the pinhole. Your back should be toward the sun to protect your eyes. To brighten the image, simply move the back piece of cardboard closer to the pinhole. To see it larger, move the back cardboard farther away. Do not make the pinhole larger. It will only distort the crescent sun.

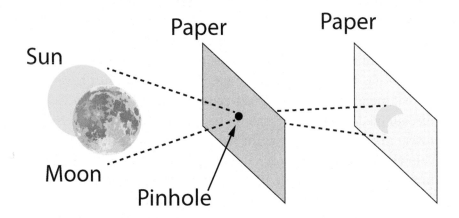

Welding Goggles

Welding goggles that have a rating of fourteen or higher are another useful eclipse viewing tool. The goggles can be used to view the solar eclipse directly. Do not use the goggles to look through binoculars or telescopes, as the goggles could potentially shatter due to intense direct heat. Avoid long periods of gazing with the goggles. Look away every so often. Give your eyes a break.

Solar Filters for Telescopes

The ONLY safe way to view solar eclipses using telescopes or binoculars is to use solar filters. The filters are coated with metal to diminish the full intensity of the sun. Although the filters can be expensive, it is better to purchase a quality filter rather than an inexpensive one that could shatter or melt from the heat.

The filters attach to the front of the telescope for easy viewing. Remember to give your telescope cooling breaks. Rapid heating can damage your equipment with or without filters attached.

WATCH OUT FOR UNSAFE FILTERS

There are several myths surrounding solar filters for eclipse viewing. In order for filters to be safe, they must be specially designed for looking at a solar eclipse. The following are all unsafe for eclipse viewing and can lead to retinal damage: developed colored or chromogenic film, black-and-white negatives such as X-rays, CDs with aluminum, smoked glass, floppy disk covers, black-and-white film with no silver, sunglasses, or polarizing films.

Some online articles state that using developed black-and-white film is safe. Those articles fail to mention the film must have a layer of real metallic silver to protect your eyes. Using developed film is discouraged. You cannot ensure the quality of the film. Feeling no discomfort while looking at the partial solar eclipse does NOT mean your eyes are protected. Retinal damage can occur with zero pain due to the retinas having no pain receptors. Please be careful. Only use protective glasses certified for viewing the eclipse.

VIEWING WITH BINOCULARS

When viewing the eclipse with binoculars, it is important to use solar filters on both lenses until totality. Only then is it safe to remove the filter. As the sun becomes visible after totality, replace the filters for safe viewing. Protect your pupils. Remember to give your binoculars a cool-down break between viewings. They can overheat rapidly from being pointed directly at the sun even with filters attached.

PLANNING AHEAD

There are many things to keep in mind when viewing a total eclipse. It is important to plan ahead to get the most out of this extraordinary experience.

UNDERSTANDING SUN POSITION

All compass bearings in this book are true north. All compasses point to Earth's magnetic north. The difference between these two measurements is called magnetic declination. The magnetic declination for South Carolina in August 2017 is:

7° 37' W ± 0° 21' (for Columbia)

Subtract the declination from the azimuth bearing as given in the text, and set your compass to that direction.

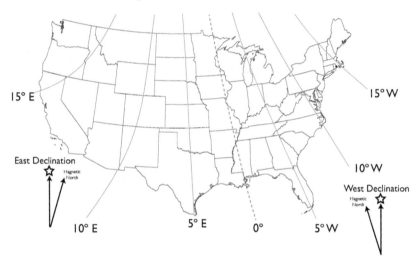

If you purchase a compass with a built-in declination adjustment, you can change the setting once and eliminate the calculations. The Suunto M-3G compass has this correction. A compass with a sighting mirror or wire will help you make a more accurate azimuth sighting.

The Suunto M-3G also has an inclinometer. This allows you to measure the elevation of any object above the horizon. Use this to figure out how high the sun will be above your position.

You can also use a smartphone inclinometer and compass for this purpose. Make sure to calibrate your smartphone's compass before every use, otherwise it might indicate the wrong bearing. Set the smartphone compass for true north to match the book. Understand the compass prior to August 21. There will be little time to guess or search on Google. Smartphone and GPS compasses are "sticky." Their compasses don't swing as freely as a magnetic compass does.

The author has used his magnetic compass for azimuth measurements and a smartphone to measure elevation. Combining these two tools will allow you to make the best sightings possible.

Outdoor sporting goods stores in most towns and cities carry compasses. We recommend purchasing a good compass in your hometown. Take the time to learn how to use it before the day of the eclipse. You do not want to struggle with orienteering basics under pressure.

SUN AZIMUTH

Azimuth is the compass angle along the horizon, with 0° corresponding to north, and increasing in a clockwise direction. 90° is east, 180° is south, and 270° is west.

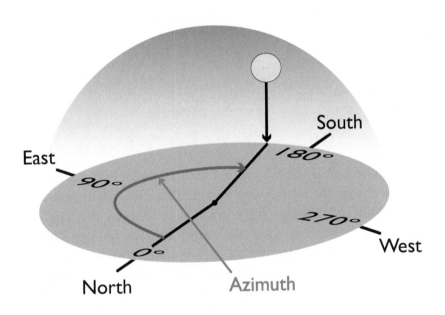

SUN ELEVATION

Altitude is the sun's angle up from the horizon. A 0° altitude means exactly on the horizon and 90° means "straight up."

Using the sun azimuth and elevation data, you can predict the position of the sun at any given time. Positions given in this book coincide with the time of eclipse totality unless otherwise noted.

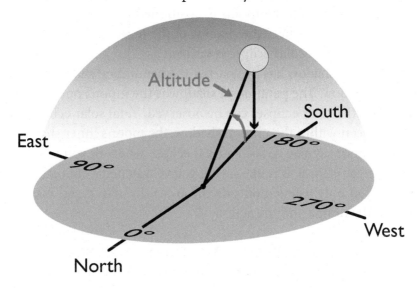

ECLIPSE DATA FOR SELECT SOUTH CAROLINA LOCATIONS

LOCATION	TOTALITY START (EDT)
ANDERSON	2:37:47PM
CAYCE	2:41:47PM
CHARLESTON	2:46:21PM
COLUMBIA	2:41:49PM
EASLEY	2:37:29PM
GREENVILLE	2:28:01PM

ECLIPSE PHOTOGRAPHY

Photographing an eclipse is an exciting challenge, as the moon's shadow moves near 2,000MPH. There is an element of danger and the pressure of time. Looking at the unfiltered sun through a camera can permanently damage your vision and your camera. If you are unsure, just enjoy the eclipse with specially designed glasses. Keep a solar filter on your lens during the eclipse and remove for the duration of totality!

PARTIAL VS. TOTAL SOLAR ECLIPSE

To successfully and safely photograph a partial and total eclipse, it is important to understand the difference between the two. A solar eclipse occurs when the moon is positioned between the sun and Earth. The region where the shadow of the moon falls upon Earth's surface is where a solar eclipse is visible.

The moon's shadow has two parts—the penumbral shadow and the umbral shadow. The penumbral shadow is the moon's outer shadow where partial solar eclipses can be observed. Total solar eclipses can only be seen within the umbral shadow, the moon's inner shadow.

You cannot say you've seen a total eclipse when all you saw was a partial solar eclipse. It is like saying you've watched a concert, but in reality, you only listened outside the arena. In both cases, you have missed the drama and the action.

PHOTOGRAPHING A PARTIAL AND TOTAL SOLAR ECLIPSE

Aside from the region where the outer shadow of the moon is cast, a partial solar eclipse is also visible before a total solar eclipse within

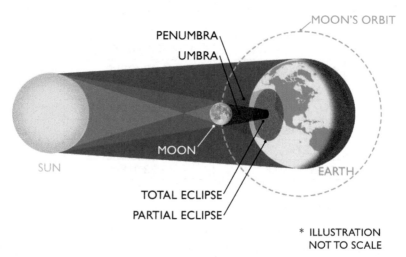

* ILLUSTRATION
NOT TO SCALE

the inner shadow region. In both cases, it is imperative to use a solar filter on the lens for both photography and safety reasons. This is the only difference between taking a partial eclipse and a total eclipse photograph of the sun.

To photograph a total solar eclipse, you must be within the Path of Totality, the surface of the Earth within the moon's umbral shadow.

THE CHALLENGE

A total solar eclipse only lasts for a couple of minutes. It is brief, but the scenario it brings is unforgettable. Seeing the radiant sun slowly being covered by darkness gives the spectator a high level of anticipation and indescribable excitement. Once the moon completely covers the sun's radiance, the corona is finally visible. In the darkness, the sun's corona shines, capturing the crowd's full attention. Watching this phenomenon is a breathtaking experience.

Amidst all the noise, cheering, and excitement, you have less than 150 seconds to take a perfect photograph. The key to this is planning. You need to plan, practice, and perfect what you will do when the big moment arrives because there is no replay. The pressure is enormous. You only have two minutes to capture the totality and the sun's corona using different exposures.

PLAN, PRACTICE, PERFECT

It is important to practice photographing before the actual phenomenon arrives. Test your chosen imaging setup for flaws. Rehearse over and over until your body remembers what you will do from the moment you arrive at your chosen spot to the moment you pack up and leave the area.

You will discover potential problems regarding vibrations and focus that you can address immediately. This minimizes the variables that might affect your photographs at the most critical moment.

It's common for experienced eclipse chasers to lose track of what they plan to do. Write down what you expect to do. Practice it time and again. Play annoying, distracting music while you practice. Try photographing in the worst weather possible. Do anything you can to practice under pressure. Eclipse day is not the time to practice.

Once the sun is completely covered, don't just take photographs. Capture the experience and the image of the total solar eclipse in your mind as well. Set up cameras around you to record not just the total solar eclipse but also the excitement and reaction of the crowd.

Eclipse Photography Gear

What do you need to photograph the total eclipse? There are only a few pieces of equipment that you'll need. Preparing to photograph an eclipse successfully takes time. Not only do you have to be skilled and have the right gear, you have to be in the correct place.

Basic Eclipse Photography Equipment

- Solar viewing glasses
- Lens solar filter
- Minimum 300mm lens
- Stable tripod that can be tilted to 60° vertical
- High-resolution DSLR
- Spare batteries for everything
- Secondary camera to photograph people, the horizon, etc.
- Remote cable or wireless release

Additional Items

- Video camera
- Video camera tripod
- Quality pair of binoculars
- Solar filters for each binocular lens
- Photo editing software

Equipment to Prepare Before the Big Day

A. Solar viewing glasses

You need a pair of solar viewing glasses as the eclipse approaches.

B. Solar Filter

Partial and total eclipse photography is different from normal photography. Even if only 1% of the sun's surface is visible, it is still approximately 10,000 times brighter than the moon. Before totality, use a solar filter on your lens. Do not look at the sun with your eyes. It can cause irreparable damage to your retinas.

DO NOT leave your camera pointed at the sun without a solar filter attached. The sun will melt the inside of your camera. Think of a magnifying glass used to torch ants and multiply that by one hundred.

C. Lens

To capture the corona's majesty, you need to use a telescope or a telephoto lens. The best focal length, which will give you a large image of the sun's disk, is 400mm and above. You don't want to waste all your efforts by bringing home a small dot where the black disk and majestic corona are supposed to be.

D. Tripod

Bring a stable enough tripod to support your camera properly to avoid unsteady shots and repeated adjustments. Either will ruin your photos. It also needs to be portable in case you need to change locations for a better shot.

E. Camera

You need to remember to set your camera to its highest resolution to capture all the details. Set your camera to:

- 14-bit RAW is ideal, otherwise
- JPG, Fine compression, Maximum resolution

Bracket your exposures. Shoot at various shutter speeds to capture different brightnesses in the corona. Note that stopping your lens all the way down may not result in the sharpest images.

Choose the lowest possible ISO for the best quality while maintaining a high shutter speed to prevent blurred shots. Set your camera to manual. Do not use AUTO ISO. Your camera will be fooled. The night before, test the focus position of your lens using a bright star or the moon.

Constantly double-check your focus. Be paranoid about this. You can deal with a grainy picture. No amount of Photoshop will fix a blurry, out-of-focus picture.

F. Batteries

Remember to bring fresh batteries! Make sure that you have enough power to capture the most important moments. Swap in fresh batteries thirty minutes before totality.

G. Remote release

Use a wired or wireless remote release to fire the camera's shutter. This will reduce the amount of camera vibration.

H. Video Camera

Run a video camera of yourself. Capture all the things you say and do during the totality. You'll be amazed at your reaction.

I. Photo editing software

You will need quality photo editing software to process your eclipse images. Adobe Lightroom and Photoshop are excellent programs to extract the most out of your images. Become well versed in how to use them at least a month before the eclipse.

J. Smartphone applications

The following smartphone applications will aid in your photography planning: Wunderground, Skyview, Photographer's Ephemeris, Sunrise and Sunset Calculator, SunCalc, and Sun Surveyor among others.

CAMERA PHONES

Smartphone cameras are useful for many things but not eclipse photography. An iPhone 6 camera has a 63° horizontal field of view and is 3264 pixels across. If you attempt to photograph the eclipse, the sun will be a measly 30-40 pixels wide depending on the phone. Digital pinch zoom won't help here. If you want *National Geographic* images, you'll need a serious camera and lens, far beyond any smartphone.

Consider instead using a smartphone to run a time-lapse of the entire event. The sun will be minuscule when shot on a smartphone. Think of something else exciting and interesting do to with it. Purchase a Gorilla Pod, inexpensive tripod, or selfie stick and mount the smartphone somewhere unique.

Also, partial and total eclipse light is strange and ethereal. Consider using that light to take unique pictures of things and people. It's rare and you may have something no one else does.

Focal Length & the Size of Sun

The size of the sun in a photo depends on the lens focal length. A 300mm lens is the recommended minimum on a full-frame (FF) DSLR. Lenses up to this size are relatively inexpensive. For more magnification, use an APS-C (crop) size sensor. Cameras with these sensors provide an advantage by capturing a larger sun.

For the same focal length, an APS-C sensor will provide a greater apparent magnification of any object. As a consequence, a shorter, less expensive lens can be used to capture the same size sun.

The below figure shows the size of the sun on a camera sensor at various focal lengths. As can be seen with the 200mm lens, the sun is quite small. On a full-frame camera at 200mm, the sun will be 371 pixels wide on a Nikon D810, a 36-megapixel body. A lower resolution FF camera will result in an even smaller sun.

Printing a 24-inch image shot on a Nikon D810 with a 200mm lens at a standard 300 pixels per inch results in a small sun. On this size paper, the sun will be a miserly 1.25 inches wide!

Photographing the eclipse with a lens shorter than 300mm will leave you with little to work with. Using a 400mm lens and printing a 24-inch print will result in a 2.5-inch-wide sun. For as massive as the sun is, it is a challenge to take a photograph with the sun of any meaningful size.

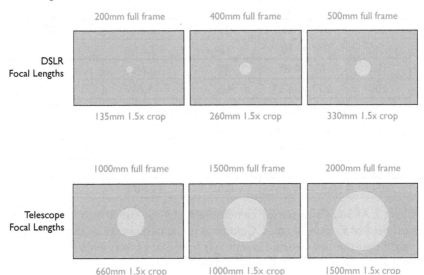

FOCAL LENGTH	FOV FULL FRAME	FF VERT. ANGLE	% OF FF	SUN PIXEL SIZE
14	104° X 81°	81°	0.7%	32.1
20	84° X 62°	62°	0.9%	41.9
28	65° X 46°	46°	1.2%	56.5
35	54° X 38°	38°	1.4%	68.5
50	40° X 27°	27°	2.0%	96.4
105	19° X 13°	13°	4.1%	200.2
200	10° X 7°	7°	7.6%	371.9
400	5° X 3.4°	3.4°	15.6%	765.6
500	4° X 2.7°	2.7°	19.6%	964.2
1000	2° X 1.3°	1.3°	40.8%	2002.5
1500	1.4° X 0.9°	0.9°	58.9%	2892.6
2000	1° X 0.68°	0.68°	77.9%	3828.4

Chart 1: Full-frame camera field of view. The 3rd column is the vertical field of view in degrees. Column 4 is the percentage of the total sensor height that the sun covers. Column 5 is how many pixels wide the sun will be on a 36MP Nikon D810. (Values are estimates)

FOCAL LENGTH	FOV CROP	CROP VERT DEG	% OF CROP	SUN PIXEL SIZE
14	80° X 58°	58°	0.9%	33.9
20	61° X 43°	43°	1.2%	45.8
28	45° X 31°	31°	1.7%	63.5
35	37° X 25°	25°	2.1%	78.7
50	26° X 18°	18°	2.9%	109.3
105	13° X 8°	8°	6.6%	245.9
200	6.7° X 4.5°	4.5°	11.8%	437.2
400	3.4° X 2°	2°	26.5%	983.7
500	2.7° X 1.8	1.8°	29.4%	1093.0
1000	1.3° X 0.9°	0.9°	58.9%	2186.0
1500	0.9° X 0.6°	0.6°	88.3%	3278.9
2000	0.6° X 0.45°	0.5°	117.8%	4371.9

Chart 2: APS-C Crop sensor camera field of view. The 3rd column is the vertical field of view in degrees. Column 4 is the percentage of the total sensor height that the sun covers. Column 5 is how many pixels wide the sun will be on a 12mp Nikon D300s. (Values are estimates)

The big challenge is the cost of the lens. Lenses longer than 300mm are expensive. They also require heavier tripods and specialized tripod heads. The 70-300mm lenses from Nikon, Canon, Tamron, and others are relatively affordable options. It is worth spending time at a local camera shop to try different lenses. Long focal-length lenses are a significant investment, especially for a single event.

To achieve a large eclipse image, you will need a long focal-length lens, ideally at least 400mm. A standard 70-300mm lens set to 300mm will show a small sun. At 500mm, the sun image becomes larger and covers more of the sensor area. The corona will take up a significant portion of the frame. By 1000mm, the corona will exceed the capture area on a full-frame sensor. See the picture on page thirty-seven for sun size simulations for different focal lengths.

SUGGESTED EXPOSURES

To photograph the partial eclipse, the camera must have a solar filter attached. If not, the intense light from the sun may damage (fry) the inside of your camera. This has happened to the author. The exposure depends on the density (darkness) of the solar filter used.

As a starting point, set the camera to ISO 100, f/8, and with the solar filter on, try an exposure of 1/4000. Make adjustments based on the histogram and highlight warning.

Turn on the highlight warning in your camera. This feature is commonly called "blinkies." This warning will help you detect if the image is overexposed or not.

Once the Baily's Beads, prominences, and corona become visible, there will only be 2.25 minutes to take bracketed shots. It will take at least eleven shots to capture the various areas of the sun's corona. The brightness varies considerably. No commercially available camera can capture the incredible dynamic range of the different portions of the delicate corona. This requires taking multiple photographs and digitally combining them afterward.

During totality, try these exposure times at ISO 100 and f/8:

1/4000, 1/2000, 1/1000, 1/250, 1/60, 1/30, 1/15, 1/4, 1/2, 1 sec, and 4 sec.

PHOTOGRAPHY TIME

Set the camera to full-stop adjustments. It will reduce the time spent fiddling. As an example, the author tried the above shot sequence, adjusting the shutter speed as fast as possible.

It took thirty-three seconds to shoot the above 11 shots using 1/3-stop increments. This was without adjusting composition, focus, or anything else but the shutter speed. When the camera was set to full stop increments, it only took twenty-two seconds to step through the same shutter speed sequence.

Assuming the totality lasts less than two minutes, only four shot sequences could be made using 1/3-stop increments. Yet six shot sequences could be made when the camera was set to full stop steps. Zero time was spent looking at the back LCD to analyze highlights and the histogram.

Now add in the bare minimum time to check the highlight warning. It took sixty-three seconds to shoot and check each image using full stops. And that was without changing the composition to allow for sun movement, bumping the tripod, etc. Looking at the LCD ("chimping") consumed **half** of the totality time.

This test was done in the comfort of home under no pressure. In real world conditions, it may be possible to successfully shoot only one sequence. If you plan to capture the entire dynamic range of the totality, you must practice the sequence until you have it down cold. If you normally fumble with your camera, do not underestimate the difficulty, frustration, and stress of total eclipse photography.

Most importantly, trying to shoot this sequence allowed for zero time to simply look at the totality to enjoy the spectacle.

AVOID LAST MINUTE PURCHASES

You should purchase whatever you think you'll need to photograph the eclipse today. This event will be nothing short of massive. Remember the hot toy of the year? Multiply that frenzy by a thousand. Everyone will want to try to capture their own photo.

Do not wait until the last few weeks before the eclipse to purchase cameras, lenses, filters, tripods, viewing glasses, and associated material. Consider that the totality of the eclipse will streak from coast to coast. Everyone who wants to photograph the eclipse will order at the same time. If you wait until August to buy what you need, it's conceivable that every piece of camera equipment capable of creating a total eclipse photo will be sold out in the United States. Whether this happens or not, do not wait until midsummer to make your purchases. It may be too late.

PRACTICE

You will need to practice with your equipment. Things may go wrong that you don't anticipate. If you've never photographed a partial or total eclipse, taking quality shots is more difficult than you think. Practice shooting the sequence with a midday sun. This will tell you if you have your exposures and timing correct. Figure out what you need well in advance.

Practice photographing the full moon and stars at night. Capture the moon in full daylight. There will be six moon cycles to practice with. Astrophotography is challenging and requires practice.

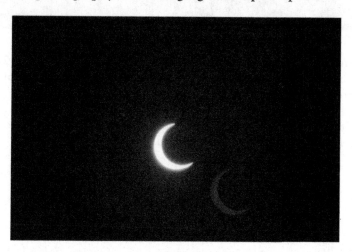

The May 20, 2012, eclipse as seen in San Diego, CA, shot with a Nikon D300s (crop sensor) with an 80-400mm lens set to ~350mm. The sun is 560 pixels wide on the 4288x2848 image.

This image is shown straight out of the camera without modification. Even with a high-quality camera and lens, photographing an eclipse is challenging. Note the haze and reflection from the overexposed sun.

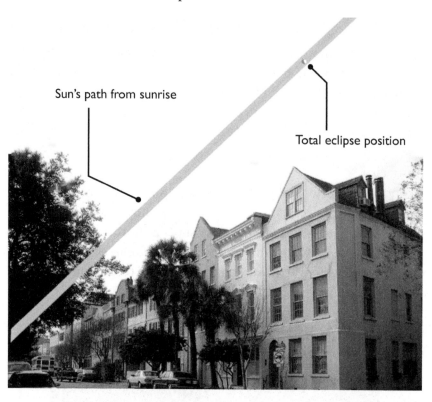

The sun will follow this path on the morning of the eclipse on August 21, 2017. Image of Charleston, South Carolina. Hopefully the sky will be free of clouds compared to what is seen in this image.

Note that this image is a simulation and approximation of where the total eclipse will appear. Refer to the eclipse position data for a more accurate location.

☉ is the symbol for the sun and first appeared in Europe during the Renaissance.
☾ is the ancient symbol for the moon.

Viewing Locations Around South Carolina

Hundreds of thousands of people will travel to and around South Carolina to view the total eclipse. The skies are partly clear, and there is a vast amount of space to view the total eclipse from.

If the weather is questionable, seek out a new location as soon as possible. If you wait until the hour before the eclipse, you may find yourself stuck in traffic, as others will be looking for clear skies. Be safe on the roadways, as drivers may be distracted.

This section contains popular, alternative, and little-known locations to watch the eclipse. As long as there are no clouds or smoke from fires, the partial eclipse will be viewable from anywhere in the state.

SUGGESTED TOTAL ECLIPSE VIEW POINTS

TOWNS AND CITIES

- Anderson
- Batesburg
- Boykin
- Cayce
- Charleston
- Columbia
- Easley
- Goose Creek
- Greenville
- Laurens
- Maudlin
- Simpsonville
- Summerville
- Sumter
- Walhalla

South Carolina Total
Eclipse Path

UNIQUE LOCATIONS

- Congaree NP
- Fort Sumter NM
- Francis Marion NF
- Lake Marion
- Lake Murray

ANDERSON

Elevation:	791 feet
Population:	27,181
Main road/hwy:	US 178

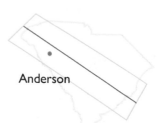

Anderson

OVERVIEW

The city of Anderson is in a county of the same name and acts as Anderson County's seat. Anderson is a large metropolis and hosts all the amenities of a big city. Hundreds of restaurants and shops call Anderson home, and there's never a dull moment. The city also has several parks where you can enjoy picnicking and jogging. While there, you may want to try some of South Carolina's famous barbecue and enjoy concerts and movie nights hosted by the city.

GETTING THERE

Drive southwest from Greenville on I-85, exit on Highway 178, and continue south until you reach Anderson.

TOTALITY DURATION

2 minutes 35 seconds

NOTES

Visit Anderson's city website for event and eclipse updates at www. cityofandersonsc.com.

Event	Time (EDT)	Altitude	Azimuth
Sunrise	6:56:00AM	0°	74°
Eclipse Start	1:08:59PM	66°	164°
Totality Start	2:37:47PM	63°	216°
Totality End	2:40:22PM	62°	217°
Eclipse End	4:03:06PM	49°	245°
Sunset	8:10:00PM	0°	285°

BATESBURG

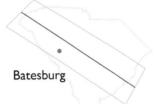

Elevation:	643 feet
Population:	5,423
Main road/hwy:	US 1

Batesburg

OVERVIEW

Batesburg is a small town in Lexington County. It's a bustling community most known for its annual poultry fair that takes place every May. There are plenty of opportunities to dine out and go shopping in town, and Batesburg is also near Lake Murray, a man-made lake where people can boat, fish, or just enjoy the sunset. If golf is your game, then you should visit the Ponderosa Country Club, an eighteen-hole regulation course with a par of seventy-two.

GETTING THERE

Drive southeast on I-20 and take exit 44 on Pond Branch Road, then continue west until you reach Batesburg.

TOTALITY DURATION

2 minutes 29 seconds

NOTES

Stop by the Batesburg-Leesville town website for eclipse event updates at www.batesburg-leesville.org.

Event	Time (EDT)	Altitude	Azimuth
Sunrise	6:52:00AM	0°	74°
Eclipse Start	1:12:02PM	67°	168°
Totality Start	2:40:57PM	62°	220°
Totality End	2:43:27PM	62°	221°
Eclipse End	4:05:45PM	48°	247°
Sunset	8:05:00PM	0°	285°

BOYKIN

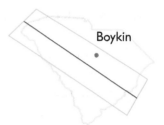

Boykin

Elevation:	164 feet
Population:	100
Main road/hwy:	US 521

OVERVIEW

Located in Kershaw County some miles south of Camden is a small rural community. Boykin is of great historical importance as the location of the final battle of the Civil War in South Carolina, as well as the place of death of the final Union soldier killed during the war. Today there are several shops and restaurants, including the Mill Pond Steakhouse, that line the shores of the mill pond created by the city's founder.

GETTING THERE

Drive northeast from Columbia on I-25 and turn south on Highway 521 to reach Boykin.

TOTALITY DURATION

1 minute 21 seconds

NOTES

Visit this South Carolina web page for updates and information on Boykin at www.sciway.net/city/boykin.html.

Event	Time (EDT)	Altitude	Azimuth
Sunrise	6:48:00AM	0°	74°
Eclipse Start	1:14:01PM	67°	172°
Totality Start	2:43:06PM	61°	223°
Totality End	2:44:27PM	61°	223°
Eclipse End	4:06:44PM	47°	248°
Sunset	8:01:00PM	0°	285°

CAYCE

Elevation:	240 feet
Population:	12,528
Main road/hwy:	US 321

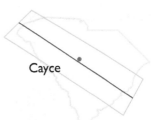

OVERVIEW

The city of Cayce is unique in that it resides in both Lexington and Richland counties and sits beside the Congaree River. Cayce is an old city with a rich history, and those who would like to learn more about that history can visit the Cayce Historical Museum. Cayce is home to many historical locations and attractions, including a replica trading post from the 1700s. This city along with all cities inside the loop in Columbia should be good places to watch the eclipse as long as the weather holds.

GETTING THERE

Cayce is a southwest suburb city of Columbia.

TOTALITY DURATION

2 minutes 32 seconds

NOTES

Visit Cayce's website for eclipse and event updates at www.cityof-cayce-sc.gov.

Event	Time (EDT)	Altitude	Azimuth
Sunrise	6:50:00AM	0°	74°
Eclipse Start	1:13:04PM	67°	170°
Totality Start	2:41:47PM	62°	221°
Totality End	2:44:20PM	61°	223°
Eclipse End	4:06:20PM	47°	248°
Sunset	8:03:00PM	0°	285°

CHARLESTON

Elevation:	20 feet
Population:	120,083
Main road/hwy:	I-26

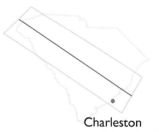

Charleston

OVERVIEW

Charleston is the oldest city in South Carolina, and it has grown tremendously since its founding. Because of how large the city is, there are several eclipse events happening concurrently, so there's something for everyone. In Charleston, you can watch the eclipse from a warship, an aquarium, a special baseball game, and even while doing yoga. There's nothing like a big city to show just how much you can do while enjoying an eclipse.

GETTING THERE

Charleston is a major city in South Carolina. Fly or drive from multiple locations to reach the city.

TOTALITY DURATION

1 minute 34 seconds

NOTES

Multiple major events are planned in the Charleston area for the total solar eclipse. Visit the Charleston website for events, lodging, and accommodations at www.charlestoncvb.com.

Event	Time (EDT)	Altitude	Azimuth
Sunrise	6:47:00AM	0°	74°
Eclipse Start	1:16:55PM	69°	176°
Totality Start	2:46:21PM	61°	227°
Totality End	2:47:55PM	61°	227°
Eclipse End	4:09:58PM	46°	251°
Sunset	7:57:00PM	0°	285°

Columbia

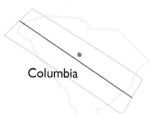

Columbia

Elevation:	292 feet
Population:	132,067
Main road/hwy:	Multiple

Overview

Columbia is the largest city in South Carolina as well as the state's capital. With a population of over a hundred thousand, there's no shortage of things to do. With all the big-city amenities, you can go out to dine at fine restaurants, take in a movie, or go to a museum or art gallery. If you're more of a nature lover, then you can go and visit the nearby mountains for hiking and camping. Multiple events are being held all around the metropolis for the total eclipse, including at the South Carolina State Museum. Visit their website for more information at scmuseum.org/eclipse.

Getting There

Columbia is a major city in South Carolina. Fly or drive from multiple locations to reach it.

Totality Duration

2 minutes 30 seconds

Notes

Visit the city's website for events and updates at www.columbiasc. gov. The Columbia total eclipse weekend website also has additional information at totaleclipsecolumbiasc.com.

Event	Time (EDT)	Altitude	Azimuth
Sunrise	6:50:00AM	0°	74°
Eclipse Start	1:13:05PM	67°	170°
Totality Start	2:41:49PM	62°	222°
Totality End	2:44:19PM	61°	223°
Eclipse End	4:06:19PM	47°	248°
Sunset	8:03:00PM	0°	285°

EASLEY

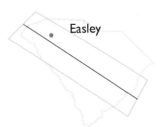

Easley

Elevation:	1,079 feet
Population:	20,549
Main road/hwy:	Hwy 93

OVERVIEW

Easley is a bustling town in Pickens County that's best known for its farmer's market. The Easley Farmer's Market takes place every Saturday at noon during the spring and summer. Every week, local farmers and craftsmen come together to sell their produce and wares to the public. Best of all, Easley holds a Music at the Market event every Friday so that people can enjoy live music while they explore the market square. This is a good way to start the eclipse weekend.

GETTING THERE

Drive west from Greenville on US 123, and continue on Highway 93 to reach Easley.

TOTALITY DURATION

2 minutes 26 seconds

NOTES

Point your web browser to the city's website for information and updates on the total eclipse at www.cityofeasley.com.

Event	Time (EDT)	Altitude	Azimuth
Sunrise	6:55:00AM	0°	74°
Eclipse Start	1:08:45PM	66°	164°
Totality Start	2:37:29PM	62°	216°
Totality End	2:39:56PM	62°	217°
Eclipse End	4:02:38PM	49°	245°
Sunset	8:10:00PM	0°	285°

GOOSE CREEK

Elevation:	46 feet
Population:	40,370
Main road/hwy:	US 78

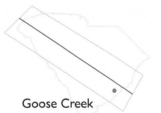

Goose Creek

OVERVIEW

Goose Creek is a modestly sized city that is quiet and friendly. Located at the heart of an industrial boom, Goose Creek is a city on the rise. The community is tightly knit, and the Community Center is a comprehensive facility that offers a fitness area, indoor track, climbing walls, and basketball courts. A new activity center is set to open in 2018 that will house two full-size gymnasiums, a fitness center, and even a preschool.

GETTING THERE

Drive north from Charleston on I-26, and exit north onto US 78 to reach the city of Goose Creek.

TOTALITY DURATION

2 minutes 10 seconds

NOTES

Visit the city of Goose Creek's website for more information and contact information for the city at www.cityofgoosecreek.com.

Event	Time (EDT)	Altitude	Azimuth
Sunrise	6:47:00AM	0°	75°
Eclipse Start	1:16:29PM	68°	175°
Totality Start	2:45:32PM	61°	226°
Totality End	2:47:42PM	61°	227°
Eclipse End	4:09:29PM	46°	250°
Sunset	7:58:00PM	0°	284°

GREENVILLE

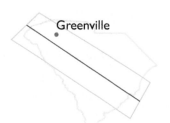

Greenville

Elevation:	965 feet
Population:	62,252
Main road/hwy:	Multiple

OVERVIEW

The Roper Mountain Science Center in Greenville will be hosting an Eclipse Extravaganza on the day of and the weekend preceding the eclipse. The extravaganza will feature full dome shows in the planetarium, informational programs on how to view the eclipse safely, and will feature special observation sites with astronomers on hand to assist visitors. If you're feeling lucky, enter the center's raffle and try to win a chance at viewing the eclipse from the observatory's telescope. If you win, you'll be the only person to view the eclipse through it.

GETTING THERE

Drive northwest from Columbia on I-26 and I-385 to reach Greenville.

TOTALITY DURATION

2 minutes 7 seconds

NOTES

Visit the Roper Mountain Science Center's website for more information on the event at www.ropermountain.org/main.asp?titleid=eclipse2017.

Event	Time (EDT)	Altitude	Azimuth
Sunrise	6:54:00AM	0°	74°
Eclipse Start	1:09:12PM	66°	165°
Totality Start	2:38:01PM	62°	216°
Totality End	2:40:09PM	62°	217°
Eclipse End	4:02:53PM	48°	245°
Sunset	8:10:00PM	0°	285°

LAURENS

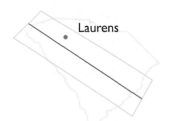

Laurens

Elevation:	610 feet
Population:	9,182
Main road/hwy:	US 76

OVERVIEW

Laurens is a town with a bustling downtown. This city has several unique shops and restaurants where people go to shop and socialize with friends. Laurens often holds festivals in their historic downtown square and offers an array of amenities to both its citizens and visitors. Thanks to a private/public partnership, Lauren's main street offers many activities such as a downtown farmer's market and wine tastings. As well, each final Friday of the month offers live music and food in the downtown square.

GETTING THERE

Drive southeast from Greenville on I-385 and exit south on US 221 to reach Laurens.

TOTALITY DURATION

2 minutes 30 seconds

NOTES

The city's website will have contact and event information for eclipse chasers. Visit it at www.cityoflaurenssc.com.

Event	Time (EDT)	Altitude	Azimuth
Sunrise	6:53:00AM	0°	74°
Eclipse Start	1:10:24PM	66°	167°
Totality Start	2:39:08PM	62°	218°
Totality End	2:41:38PM	62°	219°
Eclipse End	4:04:04PM	48°	246°
Sunset	8:08:00PM	0°	285°

Maudlin

Elevation:	951 feet
Population:	24,823
Main road/hwy:	US 276

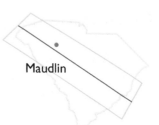

Maudlin

Overview

Located in Greenville County, Mauldin is a town that knows how to have fun. With a thriving cultural life, Mauldin offers Summer Amphitheatre Season. Every Friday people gather for a night of music, food, local wine, and craft beer. Mauldin is bursting with family-friendly activities, dozens of restaurants, and boasts over thirty acres of parks. The city's Cultural Center is home to theatre performances, festivals, and concerts.

Getting There

Drive southeast on US 276 from Greenville to reach the city of Mauldin.

Totality Duration

2 minutes 13 seconds

Notes

The city's website will have more information about eclipse events and lodging. Visit it at www.cityofmauldin.org.

Event	Time (EDT)	Altitude	Azimuth
Sunrise	6:54:00AM	0°	74°
Eclipse Start	1:09:27PM	66°	165°
Totality Start	2:38:15PM	62°	217°
Totality End	2:40:29PM	62°	218°
Eclipse End	4:03:08PM	48°	245°
Sunset	8:09:00PM	0°	285°

SIMPSONVILLE

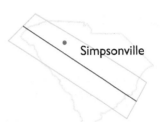

Elevation:	860 feet
Population:	21,025
Main road/hwy:	Hwy 417

OVERVIEW

There's nothing better than tasting freshly grown produce. Simpson-ville is a city that understands that. Every Saturday during the spring and summer, they host a farmer's market, one of the largest in South Carolina. The city is also host to regular fairs and even schedules movies in the park for families to enjoy. This quiet town is perfect if you want to get away from sprawling metropolises but don't want to go to the middle of nowhere either.

GETTING THERE

Drive on I-385 southeast from Greenville and exit south on High-way 417 to reach Simpsonville.

TOTALITY DURATION

2 minutes 16 seconds

NOTES

Open your web browser to the city's web page at www.simpsonville. com for more information about events and accommodations.

Event	Time (EDT)	Altitude	Azimuth
Sunrise	6:55:00AM	0°	74°
Eclipse Start	1:09:39PM	66°	166°
Totality Start	2:38:26PM	62°	217°
Totality End	2:40:42PM	62°	218°
Eclipse End	4:03:19PM	48°	245
Sunset	8:10:00PM	0°	285°

SUMMERVILLE

Elevation:	89 feet
Population:	46,974
Main road/hwy:	US-17 ALT

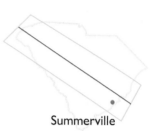

Summerville

OVERVIEW

Sometimes you can't beat a small city with lots of history, fresh produce, and tons of surrounding nature. Summerville has all three. With a weekly farmer's market, a historical downtown that hasn't changed for almost a century, and lots of trails for hiking, jogging, and biking, Summerville is the perfect spot to relax. Summerville is also steeped in South Carolina barbecue tradition. There's no shortage of southern comfort food here, and many of them are unique to Summerville.

GETTING THERE

Drive north on I-26 and exit south on US 17 ALT to reach Summerville.

TOTALITY DURATION

2 minutes 3 seconds

NOTES

Visit the city's website for more event information at www.visitsummerville.com.

Event	Time (EDT)	Altitude	Azimuth
Sunrise	6:47:00AM	0°	74°
Eclipse Start	1:16:07PM	68°	174°
Totality Start	2:45:15PM	61°	226°
Totality End	2:47:19PM	61°	227°
Eclipse End	4:09:13PM	46°	250°
Sunset	7:58:00PM	0°	285°

SUMTER

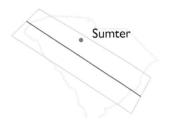

Elevation:	171 feet
Population:	40,929
Main road/hwy:	US 378

OVERVIEW

Sumter is the county seat of Sumter County and will be in the area of totality for the coming eclipse. To celebrate, Sumter is holding a total eclipse watch party at Dillon Park the day of the eclipse. The celebration kicks off with live music, kid games, and several other activities spread throughout the day. Admission is free, and there will be food, ice cream, and beverages. If you're anywhere near the area, stop by this city to see the eclipse.

GETTING THERE

Drive east from Columbia on US 378 to reach Sumter.

TOTALITY DURATION

1 minute 47 seconds

NOTES

The Sumter city website will have more event and lodging information for the eclipse weekend. Visit it at www.sumtersc.gov.

Event	Time (EDT)	Altitude	Azimuth
Sunrise	6:47:00AM	0°	74°
Eclipse Start	1:14:42PM	67°	173°
Totality Start	2:43:35PM	61°	223°
Totality End	2:45:23PM	61°	224°
Eclipse End	4:07:21PM	46°	249°
Sunset	8:00:00PM	0°	284°

WALHALLA

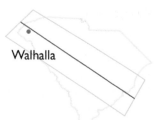

Elevation:	1,028 feet
Population:	4,218
Main road/hwy:	Hwy 28

OVERVIEW

Walhalla bills itself as an excellent access point to thousands of acres of mountain adventure in northwest South Carolina. Originally founded as a German immigrant town around 1850, the small city now boasts a population of nearly four thousand people. Due to its German heritage, Walhalla is known for its Oktoberfest celebration on the third Friday of October every year. This small, charming community will be a pleasant location to enjoy the total eclipse from as it enters South Carolina.

GETTING THERE

Drive west from Greenville on US 123, and exit northwest on Highway 28 to reach Walhalla.

TOTALITY DURATION

2 minutes 36 seconds

NOTES

Visit the city's website for more information about lodging and events for the total eclipse weekend at www.cityofwalhalla.com.

Event	Time (EDT)	Altitude	Azimuth
Sunrise	6:57:00AM	0°	74°
Eclipse Start	1:07:47PM	66°	163°
Totality Start	2:36:36PM	63°	215°
Totality End	2:39:12PM	62°	216°
Eclipse End	4:02:06PM	49°	244°
Sunset	8:12:00PM	0°	285°

Congaree National Park

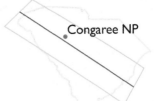

Congaree NP

Elevation:	170 feet
Main road/hwy:	Hwy 48

Overview

Congaree National Park stretches for over 25,000 miles of untainted woods. This national park is great if you enjoy fishing, hiking, or camping. Featuring dozens of breathtaking trails, the beauty of nature will surely inspire you. What's most unique about Congaree, though, is that every year, thousands of fireflies light up the forest floor in a completely synchronized display that appears only a few times during the summer. Congaree's visitor center offers information on where you can see this annual spectacle. It is one of the few national parks in the south that will enjoy the total eclipse.

Getting There

Drive southwest from Columbia on Highway 48 to Gadsden. Turn south on St. Marks Road to reach the national park.

Totality Duration

2 minutes 32 seconds

Notes

Visit the park's website for updated information at www.nps.gov/cong.

Event	Time (EDT)	Altitude	Azimuth
Sunrise	6:49:00AM	0°	74°
Eclipse Start	1:13:53PM	67°	172°
Totality Start	2:42:36PM	62°	223°
Totality End	2:45:09PM	61°	224°
Eclipse End	4:07:00PM	47°	248°
Sunset	8:02:00PM	0°	285°

Fort Sumter NM

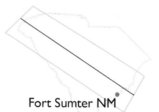

Fort Sumter NM

Elevation:	Sea level
Main road/hwy:	Boat

Overview

Fort Sumter is where the Civil War began. On 1861, Confederate artillery attacked the fort, and it surrendered more than thirty hours later. If you're a history buff, then you'll no doubt love visiting this piece of American history. Nowadays, people are working to preserve the fort and its cannons. If you want to see the eclipse, then Fort Sumter is providing viewing spots at Liberty Square for those who want a good view at this once-in-a-lifetime event.

Getting There

Take the ferry from Charleston to reach the national monument.

Totality Duration

1 minute 37 seconds

Notes

Fort Sumter's national parks' page has event updates available at www.nps.gov/fosu.

Event	Time (EDT)	Altitude	Azimuth
Sunrise	6:47:00AM	0°	75°
Eclipse Start	1:17:08PM	69°	176°
Totality Start	2:46:31PM	61°	227°
Totality End	2:48:08PM	61°	228°
Eclipse End	4:10:07PM	46°	251°
Sunset	7:57:00PM	0°	284°

FRANCIS MARION NF

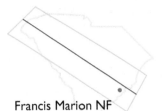

Francis Marion NF

Elevation: 47 feet
Main road/hwy: Hwy 41

OVERVIEW

Located north of Charleston, this national forest was named after Francis Marion, a hero of the Revolutionary War. With its many lakes, trails, and diverse terrains, Francis Marion National Forest is a goldmine of four-wheeling, biking, and hiking adventure. The nearby Chattooga River makes for great fishing, and there are many low-country swamps that are perfect for paddling. The Sewee Visitor and Environmental Educational Center, meanwhile, provides many programs that introduce visitors to the wonders of nature.

GETTING THERE

Drive north from Charleston on US 17 Bypass, and continue on Highway 41 north to reach the national forest.

TOTALITY DURATION

2 minutes 30 seconds

NOTES

The national forest's website has updated information on current conditions, events, access, and contacts at www.fs.usda.gov/main/scnfs/home.

Chart times for Old Joe. Location affects times.

Event	Time (EDT)	Altitude	Azimuth
Sunrise	6:46:00AM	0°	75°
Eclipse Start	1:16:55PM	68°	176°
Totality Start	2:45:42PM	61°	226°
Totality End	2:48:12PM	61°	227°
Eclipse End	4:09:40PM	46°	250°
Sunset	7:57:00PM	0°	284°

LAKE MARION

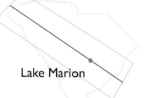

Lake Marion

| Elevation: | 89 feet |
| Main road/hwy: | Multiple |

OVERVIEW

Lake Marion is the largest lake in South Carolina and stretches along five counties. It's so large that it is often referred to as an inland sea to many locals. Because it's so large, it is the perfect spot to zoom across the surface on a boat or to just kick back and get your feet wet on a hot summer day. However, Lake Marion is best known for its huge fish. Fishing is a way of life for those who live on or near this lake. The lake lies directly on the centerline of the total eclipse.

GETTING THERE

Drive on I-26 from Charleston, and take exit 169B north on I-95 to reach Lake Marion.

TOTALITY DURATION

2 minutes 35 seconds

NOTES

Check the Lake Marion Marina Campground for more information and events at the lake at www.lakemarionresortmarina.com.

Event	Time (EDT)	Altitude	Azimuth
Sunrise	6:48:00AM	0°	75°
Eclipse Start	1:14:00PM	68°	173°
Totality Start	2:43:41PM	61°	224°
Totality End	2:46:16PM	61°	225°
Eclipse End	4:07:58PM	47°	249°
Sunset	8:00:00PM	0°	284°

LAKE MURRAY

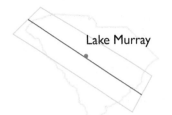

Lake Murray

Elevation:	354 feet
Main road/hwy:	US 378

OVERVIEW

Lake Murray is a reservoir that spans over fifty thousand acres with over five hundred miles of shoreline. There are dozens, perhaps hundreds, of marinas, ramps, and boat landings to get your boat into the water and start having fun. As well, several parks and recreational areas line the shores of Lake Murray, and it makes a perfect spot to enjoy a picnic or barbecue. Fishing is also a popular pastime, with fishermen reportedly catching fish weighing in at over twenty pounds.

GETTING THERE

Drive from Columbia northwest on I-26, and take exit 102A on Highway 60 west to reach Lake Murray.

TOTALITY DURATION

2 minutes 35 seconds

NOTES

Visit the lake's website for camping and visitation information at lakemurrayfun.com.

Event	Time (EDT)	Altitude	Azimuth
Sunrise	6:51:00AM	0°	74°
Eclipse Start	1:12:23PM	67°	169°
Totality Start	2:41:08PM	62°	221°
Totality End	2:43:43PM	61°	222°
Eclipse End	4:05:50PM	47°	247°
Sunset	8:04:00PM	0°	285°

Paste a meaningful picture of your eclipse experience, perhaps a photo of you and the people you watched it with. This will help you remember your family, friends, and companions.

REMEMBER THE SOUTH CAROLINA TOTAL ECLIPSE
August 21, 2017

Who was I with? _____

_____

_____

_____

_____

What did I see? _____

_____

_____

_____

_____

_____

_____

_____

What did I feel? _____

_____

_____

_____

_____

_____

_____

_____

_____

_____

_____

_____

What did the people with me think? _____

_____

_____

_____

_____

_____

_____

_____

_____

Where did I stay?_____

_____

_____

_____

_____

_____

_____

_____

_____

Enjoy other Sastrugi Press titles

2017 Total Eclipse State Series by Aaron Linsdau
    Sastrugi Press has published several state-specific guides to the 2017 total eclipse crossing over the United States. Check the Sastrugi Press website for the various state eclipse books: www.sastrugipress.com/eclipse/

Antarctic Tears by Aaron Linsdau
    What would make someone give up a high-paying career to ski alone across Antarctica to the South Pole? This inspirational true story will make readers both cheer and cry. Fighting skin-freezing temperatures, infections, and emotional breakdown, Aaron Linsdau exposes the harsh realities of the world's largest wilderness. Discover what drives someone to the brink of destruction while pursuing a dream.

Adventure One by Aaron Linsdau and Terry Williams, M.D.
    What does it take to conceptualize, plan, and enjoy your first expedition? This inspirational book contains hard-won knowledge from both authors about their experiences on expeditions around the world. The information provided in this book is useful whether you plan to climb a high peak, cross a polar plateau, or set out on a never before attempted new trek. (*Available Summer 2017*)

Lost at Windy Corner by Aaron Linsdau
    Climbing Denali is a treacherous affair. Avalanches, blinding blizzards, and crevasses have killed experienced teams. What happens when someone decides to climb the mountain solo? In this dramatic story, Aaron describes the choices made and the lessons that were learned as a result. This is more than an adventure story. It teaches defining success on your own terms in business and life. The messages will stay with you long after the end of the book. (*Available Summer 2017*)

    Visit Sastrugi Press on the web at www.sastrugipress.com to purchase the above titles in bulk. They are also available from your local bookstore or online retailers in print, e-book, or audiobook form.

Thank you for choosing Sastrugi Press.
"Turn the Page Loose"

About Aaron Linsdau

Aaron Linsdau is a polar explorer and motivational speaker. He energizes audiences with life and business lessons that stick. He delivers a message of courage by building grit and maintaining a positive attitude. Aaron teaches audiences how to eat two sticks of butter a day to achieve their goal. He shares how to build resilience to deal with constant pressure and adrenaline overload.

He holds the world record for the longest expedition in days from Hercules Inlet to the South Pole. Aaron is the second only American to complete the trip alone.

This solo expedition is more difficult than climbing Mount Everest with a team. Being alone dramatically increases the challenge. Aaron uses emotionally stirring stories to show how to overcome obstacles, impossible challenges, and unimaginable conditions. He relates these stories to business challenges and shows how the common person can achieve uncommon results.

Aaron collaborates with organizations to deliver the right message for the audience. He relates his experiences to business realities. Aaron loves inspiring audiences. Book Aaron for your next event today.

"Never Give Up"
Grit • Courage • Attitude • Perseverance • Resilience

Learn more about Aaron Linsdau at:
www.aaronlinsdau.com or www.ncexped.com.

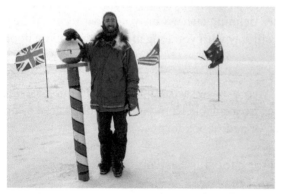

Smartphone link

Aaron at the South Pole after 82 days alone in Antarctica.

CPSIA information can be obtained
at www.ICGtesting.com
Printed in the USA
LVOW05s0420030817
543521LV00007BA/250/P